DiscoverRoo

An Imprint of Pop!
popbooksonline.com

How Animals Evolved

HOW MAMMALS EVOLVED

by Grace Hansen

WELCOME TO DiscoverRoo!

This book is filled with videos, puzzles, games, and more! Scan the QR codes* while you read, or visit the website below to make this book pop.

popbooksonline.com/ev-mammals

abdobooks.com
Published by Pop!, a division of ABDO, PO Box 398166, Minneapolis, Minnesota 55439.

Printed in the United States of America, North Mankato, Minnesota.
102023
012024

THIS BOOK CONTAINS RECYCLED MATERIALS

Cover Photo: Shutterstock
Interior Photos: Shutterstock; Getty Images; Science Source
Editor: Elizabeth Andrews
Series Designer: Laura Graphenteen

Library of Congress Control Number: 2023939056

Publisher's Cataloging-in-Publication Data
Names: Hansen, Grace, author.
Title: How mammals evolved / by Grace Hansen
Description: Minneapolis, Minnesota : Pop!, 2024 | Series: How animals evolved | Includes online resources and index
Identifiers: ISBN 9781098245450 (lib. bdg.) | ISBN 9781098246013 (ebook)
Subjects: LCSH: Evolution--Juvenile literature. | Mammals--Juvenile literature. | Mammals--Evolution--Juvenile literature. | Biology--Juvenile literature.
Classification: DDC 591.38--dc23

*Scanning QR codes requires a web-enabled smart device with a QR code reader app and a camera.

TABLE OF CONTENTS

CHAPTER 1

WHAT IS EVOLUTION?

Evolution is the scientific **theory** that all life on earth has a common ancestor that lived billions of years ago. Living things **developed** from earlier versions of that ancestor. The changes that happened took millions of years.

WATCH A VIDEO HERE!

Life forms have evolved on Earth for millions of years.

DID YOU KNOW?

Almost all scientists believe in evolution because it is testable.

Mammals are intelligent creatures.

Evolution happens because of natural selection. This process occurs when one individual of a **species** has different **traits** from others. Sometimes the differences make surviving easier for the individual. As they survive, they pass on their traits to their young. Over time, more and more individuals in the species are born with the new and helpful trait until it is normal.

Scientists believe the earliest mammals would have lived more than 200,000 years after the death of the dinosaurs.

Evolution is how fish eventually moved onto land. It is how birds began to fly and mammals started giving birth to live young. It is how humans began walking on two feet.

This ancient ancestor of elephants lived more than 34 million years ago (MYA).

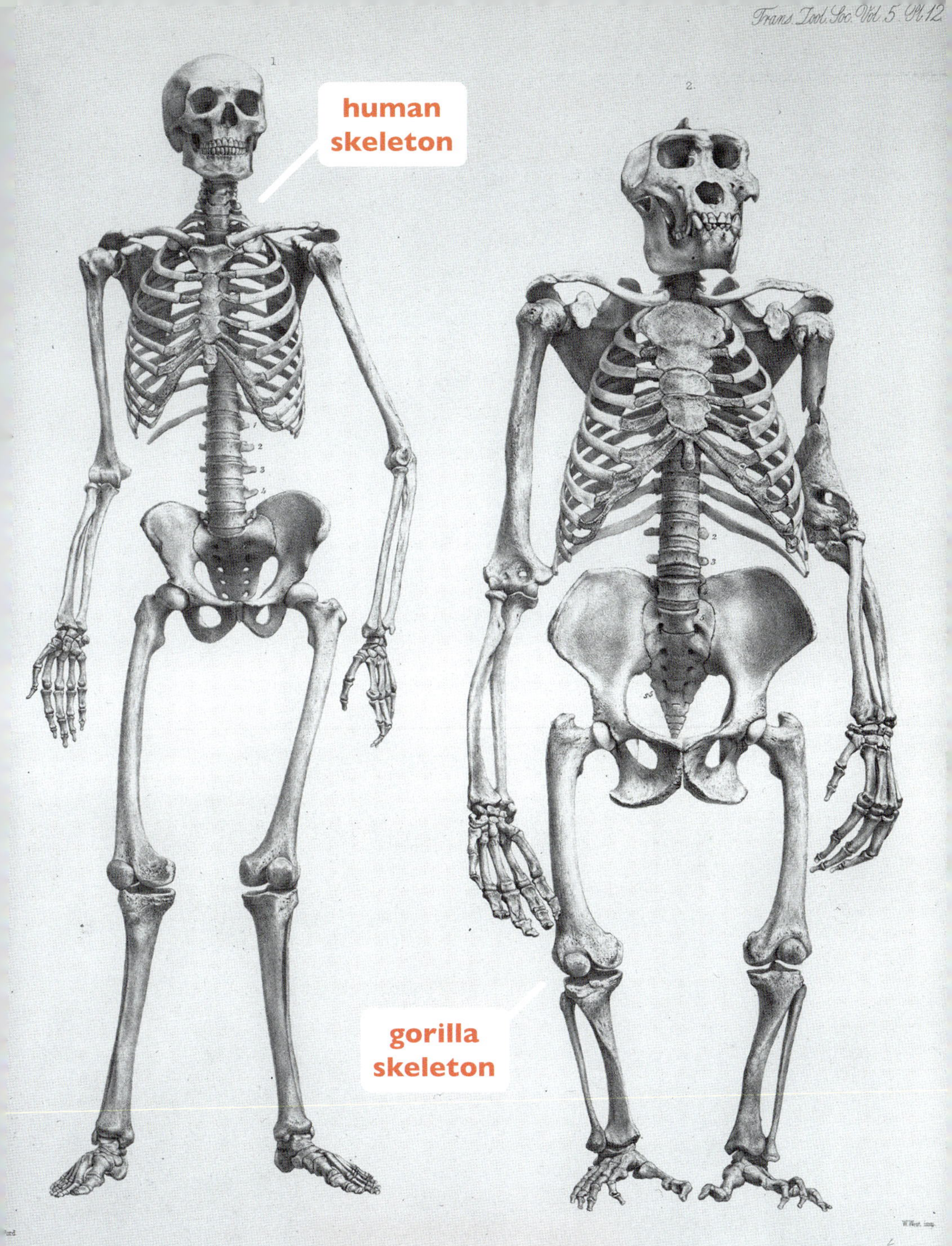

Humans and gorillas are closely related.

CHAPTER 2

WHAT ARE MAMMALS?

Mammals are animals that share certain features. Mammals get their name from a set of organs called mammary glands. When a female mammal gives birth, mammary glands produce milk.

Most young mammals are born live. They do not hatch from eggs. There are a few exceptions to this.

LEARN MORE HERE!

Young raccoons rely on their mothers for up to nine months after birth.

MORE MAMMAL TRAITS

Three tiny bones inside each middle ear space that make mammals more sensitive to sound

A unique lower jaw that gives mammals a strong bite and effective chewing

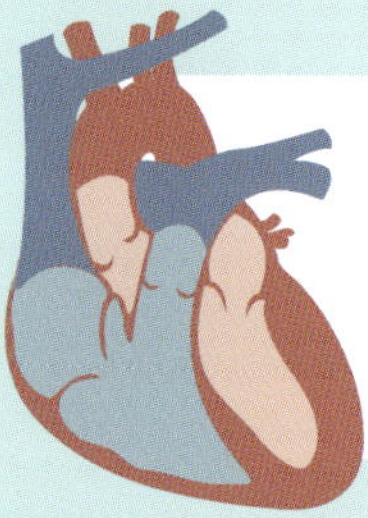

Four-chambered heart to keep oxygenated and non-oxygenated blood separate

Diaphragm to help with breathing

Well-developed brain

At some point in their life cycles, all mammals have hair. Mammals, such as cats, can use their hair to sense their surroundings. Hair also helps keep mammals warm and protects them.

Mammals can also maintain stable internal body temperatures, unlike reptiles. Reptiles warm themselves in the sun and cool themselves in the shade. Mammals keep their body temperature steady with the help of a special gland in the brain.

EGG-LAYING MAMMALS

Mammals that lay eggs are called monotremes. Only two kinds of egg-laying mammals exist today. These are the platypus and the echidna. Both these animals live in Australia.

CHAPTER 3

THE SPLIT

Scientists believe that birds, reptiles, and the first mammal-like reptiles **diverged** around 280 MYA. The **descendants** of mammal-like reptiles include all mammals. Around 200 MYA, the

EXPLORE LINKS HERE!

monotremes split off from the other mammals before the evolution of the **placenta**.

Quokkas are marsupials that live in Western Australia.

About 160 MYA, a tiny animal moved through the treetops searching for food. Its front paws grasped the branches. Its sharp teeth snagged and chewed up worms and insects. This small, furry

Juramaia sinensis ***means "Jurassic mother from China."***

animal is known today as *Juramaia sinensis. J. sinensis* weighed less than 0.5 ounces (15g). Its skull was less than an inch (2.54cm) long. It was also a placental mammal.

Scientists know that placental mammals and today's marsupials shared a common ancestor. Marsupials are mammals that give birth to immature young. The young often shelter and feed in the mother's pouch. Kangaroos and opossums are marsupials. Placental mammals have a longer period of growth in the **womb**.

DID YOU KNOW?

The young of placental mammals receive nutrients in the womb through the placenta.

A joey grows in its mother's pouch for six to nine months.

African Elephants are the largest land mammals.

In 2011, the first *J. sinensis* fossil was discovered in China. It is the oldest known fossil of a placental mammal. This fossil is evidence of when placental mammals and marsupials split apart in their evolution.

Capybaras are the largest rodents on Earth. They live in South America.

CHAPTER 4

WORLD OF MAMMALS

Today, there are more than 250 different marsupials. Most marsupials live in South America, Australia, New Zealand, and nearby Pacific Islands. Just two, the Virginia and southern opossums, live in North America. Most marsupial **species** have pouches.

COMPLETE AN ACTIVITY HERE!

Newborn Virginia opossums drink from a nipple in their mother's pouch for the first 55 days of life.

The earliest known fossil horse is called *Hyracotherium*. It lived 56 to 34 MYA in Europe and North America. It was a small animal that ate leaves. *Parahippus* lived 24 to 17 MYA. It was the first modern horse to feed on grass. It evolved to **develop** teeth that grew throughout its life. Eating tough grasses kept the teeth worn down to a proper length.

Hyracotherium (bottom) *is the most ancient ancestor of today's horses.*

Charles Darwin was born in England in 1809.

In 1859, Charles Darwin published *On the Origin of Species*. In it, Darwin stated that humans and apes share a common ancestor. Scientists believe the split between humans and other great apes occurred during the middle part of the Neogene period (23–2.6 MYA).

DID YOU KNOW?

Most scientists currently recognize 15 to 20 different species of early humans.

Dogs are canids, belonging to the family Canidae. Canidae is one of the oldest groups of meat eaters. Dog breeds look and act very differently. Still, all dogs share one common ancestor. Scientists estimate that dogs split from wolves 40,000 to 20,000 years ago.

Humans began keeping dogs as pets at least 15,000 years ago.

All cats came from *Pseudaelurus*. Various species ranged in size from a house cat to a cougar. They lived between 20 and nine MYA. In some ways, they looked like modern cats. Roaring cats, including lions and tigers, were the first line to **diverge**.

Tigers have been around for about two million years.

Humpback whales can grow up to 60 feet (18.3m) long.

DID YOU KNOW?

There are more than 6,000 species of mammals.

Whales are part of a group of animals known as cetaceans. This group also includes porpoises and dolphins. Cetaceans have a special way of breathing. Instead of breathing through nostrils, they pull in air through a blowhole.

Scientists continue to study fossils. They are working to create a more exact picture of how mammals have changed over time.

FLYING MAMMALS

Bats are the only mammals capable of flight. Their ability to fly likely occurred around 50 MYA. Bats are very good at flying and can easily catch **prey** in the air.

MAKING CONNECTIONS

TEXT-TO-SELF

What part of the evolution of mammals surprised you the most? Please explain your answer.

TEXT-TO-TEXT

Have you read any other books about animal evolution? What did the evolution of those animals have in common with mammal evolution?

TEXT-TO-WORLD

There are thousands of species of mammals in the world. With the help of an adult, look up information about your favorite mammal. Write a short paragraph explaining why it is your favorite.

GLOSSARY

diverge — to differ or deviate; to go in separate ways.

descendant — an animal that comes from a particular ancestor or group of ancestors.

develop — to grow over time.

diaphragm — a wall of muscle that separates the chest and abdomen in humans and certain animals.

placenta — an organ formed in the womb for supplying the fetus with nourishment.

prey — an animal that is hunted by other animals for food.

species — a group of living things that are very much alike.

theory — an explanation for how things work and why things happen.

trait — a quality or feature of something.

womb — the organ in the lower body of a female mammal where offspring grow before birth.

INDEX